*Science Discoveries*

# CHARLES

# DARWIN

## and Evolution

### *Steve Parker*

Chelsea House Publishers
New York • Philadelphia

**Library of Congress Cataloging-in-Publication Data**
Parker, Steve
    Charles Darwin and evolution/Steve Parker.
        p.   cm.—(Science discoveries)
    Includes bibliographical references and index.
    ISBN 0-7910-3007-5
    1. Darwin, Charles, 1809-1882—Juvenile literature.   2. Evolution
(Biology)—History—Juvenile literature.   3. Naturalists—
England—Biography—Juvenile literature.   [1. Darwin, Charles,
1809-1882.   2. Evolution.   3. Naturalists.]   I. Title.   II. Series:
Science discoveries (1994)
QH31.D2P369   1995                                      94-20656
575′.0092—dc20                                              CIP
[B]                                                         AC


Printed in China for Imago

**Acknowledgements**

**Photographic credits:**
Bridgeman Art Library, 1, 5 top, 12 top, 15 left,
    16, 21 top
Cambridge University Library, 2
Bruce Coleman Limited, 13, 14 top, 24 top
Edinburgh Photographic Library, 6
E.T. Archive, 20, 25 top
Mary Evans Picture Library, 4 bottom, 7, 10 top,
    14 center, 18, 21 bottom, 26
Robert Harding Picture Library, 8 top, 11 top,
    12 bottom, 14 bottom, 15 bottom right, 23 righ
ICCE Photolibrary, 27 left
Oxford Scientific Films Limited, 24 bottom
Science Photo Library, 10 bottom Dr. Morley
    Read, 27 right Lawrence Livermore Laboratory

Cover montage images supplied by Mary Evans
    Picture Library, Ann Ronan Picture Library,
    Rodney Shackeil, and Bridgeman Art Library

**Illustrations:**
Tony Smith, 9, 19
Rodney Shackell, 5 bottom, 8, 11 bottom, 17,
    22-23

# Contents

# Introduction

The theory of evolution is one of the most important ideas in the study of nature. It provides the basic framework for biologists who study living things. It helps us to put animals and plants into groups, and to figure out the relationships between them. It guides our thoughts on why living things look and work as they do. It makes sense of fossils. And it is a vital part of the search for the origins of life itself.

Yet less than 150 years ago, the idea of evolution was almost unknown. Many scientists in the Western world believed the literal teachings of the Bible. The Bible said that the different groups, or species, of animals and plants—from tigers to termites, trees to toadstools—had been created by God. The species were unchanged, the same now as they were on the day of creation.

A few scientists were considering the idea that species might not always stay the same. They might change, or evolve, through time. But they could not explain how this happened.

Charles Darwin, a shy English naturalist, did. He suggested that there was a struggle for existence. Plants and animals produced more offspring than could survive. Nature itself chose, or selected, which animals and plants succeeded in the struggle to live, and which died. By this continuing process of natural selection, animals and plants gradually changed, or evolved, to survive better in their surroundings. Some species died out altogether, while new ones appeared.

Darwin's ideas caused a revolution in science and society. They have shaped the thinking of scientists and other people ever since.

*Shrewsbury in Shropshire, England, in the mid-19th century. This mellow market town on the River Severn was Charles Darwin's home until he was about 16 years old, when he left for Edinburgh Medical School.*

# Chapter One
# The Early Years

Charles Robert Darwin was born in Shrewsbury, England, on February 12, 1809. His father, Robert, was a doctor, and his mother, Susannah, was daughter of the famous china maker Josiah Wedgwood. Charles' grandfather was Erasmus Darwin, who was well known in his time as a scientist with unusual ideas. Erasmus wrote on a range of subjects such as travel by air, exploring by submarine, and how plants and animals are affected by their environments.

Despite his learned father and eminent grandfather, Charles' early years in school were not outstanding. He attended Shrewsbury School, where the main lessons were in the classics, such as Latin. Many years later, he wrote: "I believe that I was considered by all my masters and by my father as a very ordinary boy, rather below the common intelligence."

Charles' grandfather Erasmus Darwin (1731–1802) wrote about scientific ideas in verse. His poem "The Botanic Garden" described the classification system of the plant kingdom. In his work Zoonomia he put forward ideas about how the environment affects living things.

## Battle with a beetle

The young Darwin did find an interest of his own—collecting animals, plants, shells, rocks and other natural objects. He read *A Natural History of Selborne* by Gilbert White, which encouraged him to go out into the countryside, observing and collecting.

One day he pulled some old bark from a tree and found two rare types of beetle, so he grabbed one in each hand. Then a third rare beetle appeared. Not wanting to lose it, Darwin put it in his mouth. However, the beetle's reaction was to squirt a foul-tasting fluid, and Darwin had to spit the beetle out!

## *From medicine to religion*

In 1825 Charles went to Edinburgh Medical School. He soon realized that medicine was not for him. He found the lectures dull, and he had to leave the operating theater because he could not stand the horrors of surgery. (This was a few years before the first painkiller, chloroform, came into use.) He gave up medicine, to the great disappointment of his father, who arranged the next-best career: In 1828 Darwin went to Cambridge University to study the Bible and become a priest. Later, he wrote: "I did not then in the least doubt the strict and literal truth of every word in the Bible."

## *Planning an expedition*

Despite being more interested in shooting partridges than attending lectures, Darwin earned a bachelor of arts degree at Cambridge in 1831. He became friendly with two professors, geologist Adam Sedgwick and botanist John Henslow, and he continued to develop his interest in rocks, fossils, animals and plants.

*Charles was supposed to study medicine in Edinburgh from 1825 to 1827. But he spent much of the time pursuing his boyhood interest in rocks, animals, plants and other aspects of nature.*

Darwin then read *Personal Narrative* by the explorer Alexander von Humboldt. Rather than become a priest right after graduation, he decided to organize a natural-history expedition to the Canary Islands. At the same time, just by chance, the Royal Navy was arranging a round-the-world survey expedition under Captain Robert Fitzroy. Fitzroy asked Professor Henslow to recommend a naturalist for the expedition.

Henslow, knowing of Darwin's interest, recommended him for the job. At first Darwin's father refused to provide the money needed, but he was eventually persuaded that this was an excellent opportunity for his son. On December 27, 1831, Charles Darwin set sail on the 235-ton HMS *Beagle*. The initial excitement was lost on Darwin, as he was seriously seasick!

*Alexander von Humboldt (1769–1859) was a famous German explorer and scientist who traveled to South America, North America and Asia. He was interested in many areas of science, from botany to astronomy (the study of planets and stars in space), and he wrote many popular books for non-scientific readers.*

## Ideas on evolution

Charles Darwin was not the first person to think about evolution. Many others had considered the subject. Their work helped him to form his ideas about *how* evolution happened.

• His grandfather Erasmus Darwin presented informal ideas about evolution in his book *Zoonomia*.
• In the year Darwin was born, the French naturalist Jean-Baptiste Lamarck put forward the first scientific ideas on the subject—"transformation," as it was then called—in his book *Philosophie Zoologique*.
• Publisher and amateur geologist Robert Chambers wrote *Vestiges of the Natural History of Creation* in 1844, although Chambers did not admit he was author. It suggested the idea of evolution and was condemned, mainly by the Church.

## Chapter Two
# *Around the World on the* Beagle

As the *Beagle* sailed across the Atlantic to South America, Darwin gradually recovered from his seasickness.

Arriving at Bahia, Brazil, Darwin's eyes were opened to the wonders of nature. His tasks were to collect specimens of plants, animals, rocks and fossils, and to make surveys and notes at each place they visited. Soon the *Beagle* was crammed full of specimen cases, which were regularly shipped back to England.

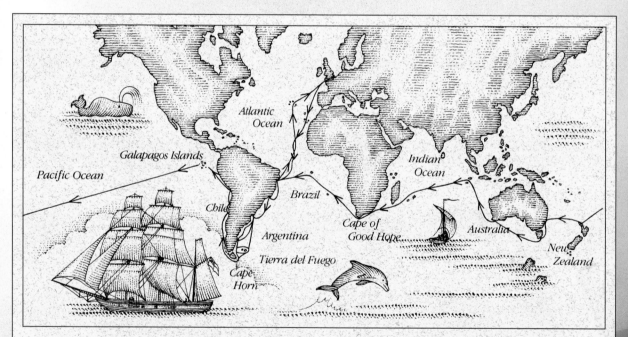

The Royal Navy commissioned Captain Robert Fitzroy (above) to make a round-the-world survey on HMS *Beagle*. They traveled to South America, rounded Cape Horn and headed up the coast to the Galápagos islands. Crossing the Pacific, the ship reached New Zealand and Australia. In each place, Darwin was busy collecting specimens of plant and animal life. From southern Australia the *Beagle* crossed the Indian Ocean, went around the Cape of Good Hope back to Brazil and then home. The whole journey took five years, 1831–1836.

*The ten-gun Beagle was about 90 feet (27.5 meters) long and weighed 235 tons. It had been launched in 1820 and had already made a long voyage to South America. It was then overhauled and refitted for its five-year survey voyage with Fitzroy and Darwin.*

*An early drawing of the bones of* Megatherium. *This huge creature, 20 feet (6 meters) long, lived from as long as a few million years ago to perhaps only several thousand years ago.*

## *Forests and fossils*

Darwin was amazed by his first walks in a tropical rain forest and wrote about "the general luxuriance of the vegetation . . . the elegance of the grasses, the novelty of the parasitical plants, the beauty of the flowers, the glossy green of the foliage . . . "

Nearby he found the huge fossilized head of an extinct giant sloth, *Megatherium.* Later in the voyage, at Port San Julian, Patagonia, he discovered fossils of another giant creature that seemed to be like a llama, but much larger.

Gradually, questions began to form in his mind. Why had certain kinds of creatures died out, like the giant sloth and the huge llama? Ordinary sloths and llamas still lived in South America. There seemed to be some sort of relationship between the fossils and the living creatures.

*The rain forest, where warmth and moisture combine to produce incredible abundance and variety in nature.*

*South American cattlemen setting up camp for the night on the pampas. Darwin noticed that different grasses grew where cattle grazed.*

## Forces for change

As the *Beagle* sailed south, Darwin began to ponder the reasons for change in the natural world. In Argentina, he noticed that the coarse grass of the pampas no longer grew in the areas where cattle had been introduced. The new grass was smaller and finer. The cattle's grazing and droppings seemed to have changed the natural pampas grass, or allowed different grasses to flourish.

At the tip of South America, in Tierra del Fuego, Captain Fitzroy took ashore three Fuegian people he had seized on a previous voyage, as hostages for a stolen boat. Darwin marveled at how the local people stood in the snow and sleet with only a few animal skins for protection. They slept on the wet ground, out in the open. He wrote: "Nature has fitted the Fuegian to the climate and productions of his miserable country." He must have been thinking about how living things (in this case humans) change or adapt to their surroundings.

### The record in the rocks

Charles Darwin collected many fossils on his voyage. At the time, fossils were known to be preserved remains of long-dead animals and plants. The accepted view was that fossils had formed after a great catastrophe, such as Noah's flood, which killed most living things.

One awkward fact was that fossils from deep layers of rock were very unfamiliar, but those from shallower layers seemed more and more similar to the species of today. The great French fossil expert Baron Georges Cuvier explained this by saying there had been a series of divine catastrophies. Each wiped out all forms of life. Then a new and improved selection of animals and plants was created to live on the Earth.

## Continual change

In *The Principles of Geology* (1830–33), Charles Lyell wrote about the principle of uniformitarianism. That theory says that processes we see today in nature, such as coastlines being worn away by the sea, or earthquakes creating huge earth movements, happened in the past too. Over a very long time, they shaped the Earth we see today. This may seem obvious now, but it was a new idea at the time.

Darwin read Lyell's book. He realized that changes in the living world, which he was observing in his travels, could also have taken place in the past. As environments changed, so might the animals and plants. Lyell's writings were very important in Darwin's thinking on evolution.

## Earthquake in Chile

The *Beagle's* surveys followed layers of rocks along a 1,200-mile stretch of South American coast, from the Rio de la Plata to Tierra del Fuego. Darwin noted that the same rocks in the south were 300 feet higher above sea level than those in the north. The entire continent seemed to have tilted. Had it moved since its formation?

The ship rounded Cape Horn and sailed up South America's west coast. On February 20, 1835, a great earthquake shook the region. Entering the port of Concepción, Chile, Darwin saw the appalling damage. He also noted that the rocks around the harbor had been lifted 2 or 3 feet by the earth's movement. Shellfish and seaweeds that were normally near the water were now high and dry. Could such catastrophic changes in the surroundings be linked to changes in plants and animals?

*An opossum from Argentina, one of the many animals Darwin described in his travels. American opossums are one of the two families of marsupial (pouched) animals that live outside Australia.*

# Chapter Three

# Evolution at the Equator

In 1835, the *Beagle* left South America and set sail across the Pacific Ocean. Some 600 miles (1,000 kilometers) from the mainland, it anchored at a group of about 13 small, rocky islands on the Equator. They were the Galápagos Islands.

Darwin was at once struck by the unusual birds, reptiles and other animals on the islands. He had never seen these particular species before—they seemed unique to the islands. Yet they had many similarities to species from the South American mainland.

## Tortoises and finches

Stranger still, each island had its own type of each animal. Most amazingly, there were giant tortoises on the islands. The ship's crew rode them like horses. The local people could tell which island a tortoise came from by the shape of its shell.

There was also a different kind of mockingbird on each island. Many of the flowers were unique to each island as well, yet similar to one another.

*Giant tortoises lived on many Pacific islands besides the Galápagos. Today they are very rare and protected by law.*

*Parts of the Galápagos are very rocky and jagged. The islands were formed by underwater volcanoes. Plant and animal life started on the islands about one million years ago.*

Darwin was especially intrigued by one group of birds, the finches. They were generally small and a drab brown color. But each species had a slightly different size and shape of beak, which meant it could tackle a certain kind of food. Darwin wrote in his notebook: "One might really fancy that from an original paucity [scarcity] of birds in the archipelago one species had been taken and modified for different ends." The idea of evolution was taking root.

*A marine iguana basking in the sun. These unique lizards dive into shallow water and eat seaweed. This one comes from Hood Island.*

*A painting of a purple-stained orchid found on the Galápagos Islands. Darwin found many varieties of orchids on the islands.*

## Darwin's finches

The 13 species of Galápagos finches live nowhere else in the world. Each species has a particular type of beak, which is suited to a certain kind of food. For example:

**1.** The large ground finch has a huge crushing beak for cracking tough seeds.
**2.** The medium ground finch has a slightly smaller but strong beak, for cracking slightly smaller hard seeds.
**3.** The warbler finch has a long, slim beak for probing into cracks and catching insects.
**4.** The small ground finch has a very small but strong beak, for cracking small, hard seeds.

Today these finches are seen as a typical example of evolution. The Galápagos Islands formed from underwater volcanoes only a few million years ago. It is thought that a few finches from South America landed there, blown by a storm. With other animals and plants already established, the finches had plenty of food. The original species evolved into many different species. Each adapted to the food source that was available on its own island.

*The Maori people lived in New Zealand long before Europeans arrived. These canoeists in Milford Sound, South Island (left), are looking for food from the sea, such as ash and shellfish. The Maori person below was pictured in 1847, a few years after Darwin's visit. His feather-edged cape shows he is the chief of his group.*

## The Pacific coral islands

The *Beagle* sailed on across the Pacific to Tahiti, where Darwin fell in love with the misty peaks, tropical plants, colorful animals and natural lifestyle of the native people.

The journey continued on toward New Zealand, and then Australia. Darwin was shocked by the terrible living conditions of the native people who were ruled over and made into slaves in their own lands by the European settlers. This seemed to support his observations from the animal world, that the stronger always took over from the weaker.

In the Indian Ocean Darwin, now a seasoned traveler and still collecting specimens by the hundred, formed a theory of how coral barrier reefs and atolls were made (see page 17).

# Chapter Four
# *Back to England*

The *Beagle* and its crew returned to Falmouth, England, on October 2, 1836. Darwin spent the next few years organizing and cataloguing his vast collections of plants, animals, rocks and fossils. He was helped by Sir Richard Owen, who was later to become one of his main opponents.

In 1839 Darwin married his cousin Emma Wedgwood. The next year his book *Journal of Researches into the Natural History and Geology of the Countries Visited During the Voyage Round the World of HMS Beagle* was a best-seller, despite its very long title. He had become a member of the Royal Society and was respected as a scientist and author. In the same year, he moved to Down House near Bromley, where he lived for the rest of his life.

*The young Emma Wedgwood, who married Darwin. Among their ten children were the botanist Sir Francis Darwin and the mathematician Sir George Darwin.*

*Down House, home to Charles Darwin and his family for more than forty years.*

## Coral creations

During Darwin's time, there were several ideas about how coral islands formed. One was that the circular coral atolls grew around the rim of a submerged volcano. Another was that the stony, cup-shaped skeletons of the tiny coral animals grew and accumulated from the seabed upward.

Darwin used his gifts for observation and sifting out the vital facts. He saw that corals grew only in the shallows, not in deep water. He reasoned that an undersea mountain, with its tip above the water, slowly sank or the sea level around it rose. The corals, trying to stay in the brightly lit shallow water, built their stony skeletons one generation upon another. As the mountain sank, the rocky reef became thicker.

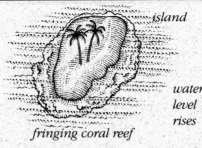

island

water level rises

*fringing coral reef*

*There are three stages in the formation of a coral atoll. Favorable conditions for its formation are shallow warm water, plenty of sunlight and nutrients.*

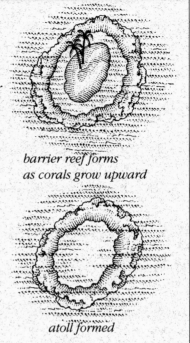

*barrier reef forms as corals grow upward*

*atoll formed*

## Work at Down House

During the 1840s and 1850s, Darwin continued his research and writing at Down House. For a time, he returned to his first love, geology.

During the voyage of the *Beagle,* the coral lagoons of the Cocos Islands in the Indian Ocean had set him thinking about how these great limestone structures formed. His book *The Structure and Distribution of Coral Reefs* came out in 1842. Two years later he published *Geological Observations on Volcanic Islands,* and after another two years *Geological Observations of South America.*

As time went on, his health began to fail. He could do only a few hours of work each day, and he took to walking in the gardens of his house and going on carriage rides. His illness was never identified, though it may have been a tropical sickness, such as Chagas' disease, caught on his voyage around the world.

## A flash of inspiration

Despite his ill health, Darwin continued his research into the idea of evolution. He was becoming more convinced that species were not fixed and immutable. They changed. He had written a short version of his ideas in 1842, but decided to collect every scrap of information he could and to write a lengthy book with masses of evidence for his theory. He even talked to pigeonkeepers about the way they bred together selected birds, to produce new kinds of pigeons. This was a form of "artificial selection."

But *how* did species change in nature? What force made them alter gradually with time? One day, while out in his carriage, Darwin came upon the idea of natural selection. It was like artificial selection, but nature did the choosing.

## "Confessing a murder"

For many years, Darwin was reluctant to publish his ideas on evolution by natural selection—that animals and plants evolved naturally. He now believed that God had not created them exactly as they were. However, most people at the time—including many scientists—still believed in the literal teachings of the Bible.

In 1844 he wrote to his close friend Joseph Hooker, director of the Royal Botanic Gardens in Kew, London. He explained his worries: "I am almost convinced (quite contrary to the opinion I started with) that species are not (it is like confessing a murder) immutable." Like the Italian scientist Galileo two centuries before, Darwin knew that speaking out against the accepted interpretations of the Bible was certain to offend and cause a storm of protest.

Darwin might never have finished this work on evolution had it not been for a letter that arrived at Down House in June 1858 from Malaya.

*Joseph Hooker (1817–1911) took over as director of Kew Gardens from his father, William. He traveled to India and the Himalayas (where he is in the picture below) and brought many plants back to England, including, in 1849, the now-familiar rhododendrons.*

## Malthus and the struggle for existence

Darwin was greatly influenced by a book called *An Essay on the Principle of Population* (1798) by Thomas Malthus, a clergyman, mathematician and economist. Malthus said that the human population could not keep increasing forever. Someday we would run out of food, living space and other things we need to survive. Then there would be a "struggle for existence," and only the strongest and fittest would survive. He identified three "evils" that reduced our numbers—war, famine and disease. Darwin took this idea and applied it to animals and plants in the natural world. He called his theory "survival of the fittest," using a phrase coined by philosopher Herbert Spencer.

*Charles Darwin at work in his study in Down House.*

# Chapter Five

# The Book that Shook the World

The letter was from another English naturalist, Alfred Wallace. Wallace knew that Darwin was interested in evolution. So with his letter he sent his summary of the theory, titled *On the Tendency of Varieties to Depart Indefinitely from the Original Type.*

Darwin was amazed. All the work he had done so patiently for the past twenty years was neatly described by Wallace. He said: "'Even his [Wallace's] terms stand as heads of my chapters.''

Fellow scientists Lyell and Hooker advised that Darwin and Wallace should have their work read as soon as possible to a scientific meeting. This happened at the Linnean Society in London, in July 1858. Then Wallace agreed that Darwin, who had gathered far more evidence to support their joint theory, should continue with the idea, while he stood aside. Darwin did so, quickly finishing his great book. It was published on November 24, 1859, and called *On the Origin of Species.*

## Alfred Wallace

Alfred Wallace traveled widely in South America and Southeast Asia, collecting specimens for museums. Like Darwin, he had been amazed by the fabulous variety of life in the tropical forests and wondered how it came to be. While resting from a fever at his camp in Southeast Asia, he remembered the same book by Malthus that Darwin knew. He had the same flash of inspiration and came upon the idea of evolution by natural selection: "On the whole, the best fitted live. From the effects of disease the most healthy escape; from enemies, the strongest, the swiftest, or the most cunning; from famine, the best hunters or those with the best digestion; and so on."

*Alfred Wallace first worked as a surveyor for the new railways in England. At the age of 25 he began to travel and collect new kinds of animals and plants, deep in the jungles of South America and Southeast Asia.*

*Charles Darwin and his new ideas became the subject of many jokes and comic pictures. As you can see from these cartoons, his theories were misunderstood.*

## *Reaction to* On the Origin of Species

The publisher of Darwin's book, John Murray, read it before printing and realized a great outcry would follow. Only 1,250 copies were printed. These sold out almost at once, so a second edition was produced.

People were indeed outraged. Darwin was denying the truth of the Bible. Scientists lined up to have their say. Former colleagues, such as the famous naturalist Philip Gosse, Richard Owen (who had helped with the *Beagle* specimens) and Adam Sedgwick (the professor at Cambridge) turned against him. In North America Louis Agassiz, professor at Harvard University and a follower of Cuvier, criticized Darwin. One clergyman called the quiet, mild-mannered Darwin "the most dangerous man in England."

But others rapidly recognized the good science in Darwin's ideas and the vast amount of evidence that supported them. The biologist Thomas Huxley defended him in England, along with Hooker and Lyell. A professor of botany at Harvard University, Asa Gray, was Darwin's great supporter in North America. Darwin himself remained at Down House and took little part in the arguments.

### *An ape for a father?*

One misunderstanding about *On the Origin of Species* concerned human evolution. Popular newspapers and cartoons said that Darwin suggested humans had descended from living apes, such as gorillas or chimps. This was not true. His only mention of the subject was: "Much light will be thrown on the origin of man and his history."

There are only two kinds of elephants alive today, the African and Asian elephants. But fossils show that in the past, there have been many other members of the elephant group, including:

(1) Moeritherium, a small animal that lived about 35 million years ago.

(2) Gomphotherium, a long-jawed animal that lived in Africa, North America and Eurasia around 20 million years ago.

(3) Platybelodon, a shovel-tusked member of the elephant group that lived 10 million years ago.

(4) The imperial mammoth, the largest of the elephant group, with tusks like

(5) the modern African elephant. The mammoth lived a million years ago. Evolution worked on the basic elephant design and fitted each kind to the conditions at the time. For example, the imperial mammoth had very long, thick hair, to keep warm and survive the cold of the last Ice Age.

## Evolution by natural selection

Evolution simply means change. Animals and plants change with time. Darwin showed that nature did the choosing. The theory of evolution is:

*Reproduction:* Parents have young that are like themselves and different from other species. Tigers have baby tigers, lions have baby lions, and so on.

*Too many young:* Not all the young can survive. Darwin calculated that in 750 years, one pair of elephants would have 19 million descendants—if they all survived.

*Variation:* not all the offspring are the same. There are slight differences in size, strength, color, or other features. New variations appear in each generation.

*Natural selection:* Life is a struggle to find food, living space, mates, and other essentials. Some features might help in this struggle, such as sharper teeth in a hunter, or more seeds in a flower. These features make a living thing better suited, or better adapted, to its surroundings. They improve its chances of surviving and breeding.

*Inheritance:* If a useful feature is inherited, the offspring of the animal or plant will have it too. It will help them to survive and leave even more offspring, who also possess the feature.

*Evolution:* Over long periods of time, and many generations, features that improve survival become more common in a species. The species changes.

*Origin of species:* Species that are best suited to the environment survive. Those that cannot adapt die out. As the environment changes, species evolve.

### What's in the Origin?

*On the Origin of Species by Means of Natural Selection, or the Preservation of Favoured Races in the Struggle for Life* is a long but very readable book. It begins by looking at "variation under domestication," including pigeons, horses and garden flowers. Then it covers variation in nature, and the problems of identifying a species. It shows how the offspring of parents are all similar, but slightly different. These slight variations might give an individual a better chance of succeeding and staying alive. Chapters 6 and 7 deal with "difficulties of the theory, and objections," Chapter 8 with animal instincts, and the later chapters with fossils and the geographical ranges of species. Darwin draws evidence from all manner of animals and plants, from mice to elephants, asparagus to furze bushes. Yet he never explains the origin of any one particular species, only how they might have evolved.

### Domestication

*Sheep were domesticated more than 10,000 years ago. They are needed for their wool, milk and meat. Above is an Australian domestic sheep, bred especially for wool and meat. Below is the barbary sheep, bred for its milk and meat. Breeders use artificial selection to produce hundreds of varieties of sheep.*

4

5

*A stag's antlers and a peacock's tail are examples of sexual selection, a form of natural selection. Females choose the male with the most impressive features for their mate, and the male offspring inherit these features. Over a long time, male deer evolved bigger antlers to impress the females. Male peacocks with the finest tails were selected by the females.*

# Chapter Six

# The Struggle for Acceptance

*On the Origin of Species* shocked and angered many people, including Darwin's own family. To accept the theory of evolution meant accepting that the Bible's version of the creation of animal and plant species could not be literally true. Many scientists struggled to believe in both. Gradually, however, the theory of evolution by natural selection gained ground, and most scientists realized that Darwin was right.

Darwin did not retire after *On the Origin of Species*. He kept up his studies and research and continued with his experiments and nature observations in and around Down House. In 1871 he published *The Descent of Man and Selection in Relation to Sex*. In this he concluded that humans are not the result of special creation but have evolved, like other animals. Their ancestors can be traced back far into prehistory. Ultimately, all living things are descended from the "filament of life" that his grandfather Erasmus had mentioned in his writings.

## Sexual selection

Natural selection says that the features of a living thing—its size, shape, color, inner organs, body chemistry, behavior, and so on—evolve to increase its chances of survival. But some features seem to reduce these chances. Wouldn't the splendid, shimmering tail feathers of a peacock get caught in the undergrowth, or be noticed by predators, and be a hindrance?

Darwin explained this by the process of sexual selection. Females are attracted to and mate with the peacock with the most impressive tail feathers. This male will pass on the feature to his offspring. It is important for individuals to survive, but it is also vital that they have offspring.

*Charles Darwin became very famous in his later years. He was visited by the foremost scientists, but he usually preferred to live a quiet life with his wife and family at Down House.*

## *The final years*

In the 1870s Darwin's health improved, and in 1877 he was awarded a special degree by Cambridge University. He continued to write books: about insect-eating plants, on how plants grow and move, and about how earthworms encourage decay and enrich the soil—they are "nature's first gardeners."

After a mild heart attack in December 1881, Darwin died peacefully at Down House on April 19, 1882, at age seventy-three.

The storm of protest over *On the Origin of Species* had died away. Charles Darwin had become a national figure and one of the best-known scientific names of all time. He was laid to rest at Westminster Abbey, London, next to the great Isaac Newton. The funeral was attended by dozens of politicians, inventors, explorers, scientists and artists, along with members of the scientific societies of many countries.

## Chapter Seven
# After Darwin

Darwin's work explained why animals and plants have the features they do. It made sense of the scheme for grouping and classifying species devised mainly by the Swedish naturalist Carolus Linnaeus. Certain groups of species are similar because they are closely related, having evolved from the same ancestor.

It also explained fossils. They represented long-dead animals and plants, most of which had lost the struggle and become extinct. However, some had evolved into different species, which were successful for a while. Fossils traced these patterns of evolution through time.

## Practical benefits

The theory of evolution by natural selection also had many practical results. It spurred research and field-work, and gave scientists a basis on which to design their experiments and make observations. When looking at any feature of a plant or animal, the biologist asks: "What is it for? How does it help survival and reproduction?"

## Mendel and inheritance

One problem, which Darwin always admitted, was that no one knew *how* features were passed on from parents to young. Why did offspring have some features of their parents but not others? And why did offspring vary slightly?

At the time when *On the Origin of Species* was causing such fuss, an Austrian monk was experimenting with peas in a peaceful monastery garden. His name was Gregor Mendel, and his work was the beginning of the modern science of genetics. It explains how certain features are inherited, controlled by what we now call genes. It shows how genes become shuffled and changed (mutated) from one generation to the next. The importance of Mendel's work was recognized only in about 1900. It solved many puzzles about inheritance, and filled some gaps in the theory of evolution.

*Gregor Mendel carried out many tests on peas. He described how features such as the yellow or green color, or smooth or wrinkled skin, were passed on from one generation of pea plants to the next, year after year.*

## Neo-Darwinism

In Darwin's day, the theory of evolution by natural selection became known as Darwinism. The more complete theory of evolution we have today is sometimes called neo-Darwinism (new Darwinism). It combines natural selection with the theory of heredity developed from Mendel's work, and with more recent developments such as the nature of mutation and the discovery of DNA.

*A greatly enlarged image of a tiny portion of DNA, the genetic substance that passes features from parent to offspring.*

*The Charles Darwin scientific research buildings in the Galápagos were named in memory of the island's most famous visitor. The islands are now part of Ecuador.*

### Evolution in fits and starts

Evolution generally takes a very long time. It happens over hundreds of generations, and thousands or millions of years. Many scientists assumed it was a gradual and continuing process.

In the 1970s, a newer idea said that evolution may often happen in fits and starts. Species stay the same for a very long time. Then they change rapidly, in a burst of evolution over a relatively short time, before settling down again. This theory is called punctuated equilibrium. It may be important in some groups of animals or plants, and it is still being discussed.

# The World in Darwin's Time

| | 1800-1825 | 1826-1850 |
|---|---|---|
| **Science** | **1801** Jean-Baptiste Lamarck publishes early ideas on evolution<br><br>**1809** Charles Darwin is born<br><br>**1820** Hans Oersted discovers the connection between electricity and magnetism | **1841** Richard Owen invents the term "dinosaur"<br><br>**1844** Robert Chambers is the unnamed author of *Vestiges of the Natural History of Creation*<br><br>**1846** The planet Neptune is discovered |
| **Western Expansion and Exploration** | **1820** Antarctica sighted for the first time, separately by a Russian, an American and an Englishman | **1842** Britain wins Hong Kong from China after Opium War<br><br>**1848** Henry Bates sets out for the Amazon, collecting about 14,000 species of insects in seven years<br><br>**1849** American gold rush in California |
| **Politics** | **1804** Napoleon becomes Emperor of France<br><br>**1815** Napoleon is defeated at Battle of Waterloo<br><br>**1821** Central America gains its freedom from Spain<br><br>**1825** Indonesians rebel against Dutch in Java War | **1830** Khartoum, founded by Mehemet Ali, Egyptian viceroy, becomes capital of Sudan<br><br>**1837** Victoria crowned Queen of England |
| **Arts** | **1816** Lord Byron writes "*Darkness*," following the massive eruption of Mount Tambora in Indonesia that darkens the world's skies for a year<br><br>**1823** Ludwig von Beethoven completes his Ninth Symphony | **1831** Katsushika Hokusai completes his series of landscape paintings *Thirty-Six Views of Mount Fuji*<br><br>**1833** Boston Academy of Music founded<br><br>**1843** Charles Dickens writes *A Christmas Carol* |

**1859** *On the Origin of Species* is published

**1860** Mendel experiments with peas; beginning of science of genetics

**1869** Dmitri Mendeleev produces the first periodic table of chemical elements

**1876** Alexander Graham Bell takes out a patent on his invention of the telephone

**1879** Louis Pasteur makes discoveries that will lead to vaccination against many diseases

**1882** Charles Darwin dies

**1854** American Matthew Perry opens Japan to trade

**1860** John Speke finds the source of the "White Nile"

**1871** Henry Stanley meets David Livingstone on the banks of Lake Tanganyika, Africa

**1879** Adolf Nordenskjöld sails the Northeast Passage, along the Arctic coasts of Europe and Asia

**1898** U.S. annexes Hawaii

**1861** American Civil War begins

**1865** Abraham Lincoln is assassinated

**1867** Karl Marx publishes *Das Kapital*

**1870** Otto von Bismarck unites Germany

**1882** Politicians stop the building of the first Channel Tunnel between England and France

**1898** U.S. battleship *Maine* blows up off the coast of Cuba; beginning of Spanish-American War

**1851** The Crystal Palace is built for London's Great Exhibition

**1865** *From the Earth to the Moon* by Jules Verne is published

**1874** Richard Wagner finishes his mammoth series of operas, *The Ring of the Nibelung*

**1874** First large-scale exhibition of Impressionist paintings is held in Paris

**1879** Prehistoric cave paintings, over 10,000 years old, discovered in Altamira, Spain

**1885** *Huckleberry Finn* by Mark Twain is published in the U.S.

**1899** Scott Joplin publishes his ragtime composition "Maple Leaf Rag"

# Glossary

**archipelago:** a group of islands

**artificial:** Caused or made by humans; not occurring in nature. In artificial selection, people select the animals they wish to breed together, to produce new varieties. In natural selection, nature does the choosing.

**atoll:** a low, ring-shaped coral island in the ocean with a lagoon in the center.

**botany:** the branch of science that studies plants. People who study botany are called botanists.

**Chagas' disease:** an illness from the tropical regions of South America. It is caused by a microscopic parasite that multiplies in the body, and it is spread in the droppings of certain insects. A similar parasite causes sleeping sickness in Africa.

**classify:** to organize things into groups based on a system or principle.

**chromosome:** a microscopic rod-shaped body found in a cell's nucleus that carries information about hereditary characteristics.

**digestion:** the breakdown of food by enzymes and bacteria into a form that can be absorbed by the body.

**DNA:** deoxyribonucleic acid, a compound that forms chromosomes, which determine what characteristics are passed on from parents to their offspring.

**domestication:** the process by which wild animals are made tame and useful to humans.

**evolution:** gradual change over time, especially in living things.

**famine:** large-scale shortage of food, during which many people become ill or die of starvation.

**filament of life:** a formerly popular name for the first kinds of life to appear on Earth, thought to be shapeless microscopic threads.

**fossils:** remains (or imprints) in rock of bones, teeth, shells, bark and other long-dead remains of once-living things.

**furze:** another name for prickly gorse, which has green spiky leaves and yellow flowers.

**gene:** the specific portion of a chromosome that determines one particular characteristic of a living thing. For example, inheriting a gene for brown eyes means an offspring will also have brown eyes.

**genetics:** the branch of biology that studies how features are passed from parents to offspring, and also how these features change over generations.

**geology:** the branch of science that studies the structure and history of Earth. People who study geology are called geologists.

**immutable:** fixed and unable to change.

**inherit:** to receive features, such as color or size, from ancestors.

**naturalist:** a person who studies animals, plants, rocks and other aspects of nature, especially by direct observation in the field.

**pampas:** the rolling grasslands of South America.

**parasite:** a plant or animal that depends on another organism for food or protection, but offers nothing in return.

**rain forest:** a dense evergreen forest that has an abundant rainfall throughout the year.

**reef:** a chain of rocks or rocklike material at or just below the water's surface. A reef is usually built up over thousands of years from the stony skeletons of millions of tiny coral animals.

**species:** a group of animals or plants that have been classified together because they can breed and produce fertile offspring.

**survey:** the study and mapping of an area to discover and record the hills, valleys, coasts, water and islands; the kinds of soils and rocks on the surface; and the plants and animals living there.

# Index

STEVE PARKER has written more than 40 books for children, including several volumes of the Eyewitness series. He has a bachelor of science degree in zoology and is a member of the Zoological Society of London.

J
B
Darwin
P

Parker, Steve.

Charles Darwin and
evolution.

| DATE | | | |
|---|---|---|---|
| | | | |
| | | | |
| | | | |
| | | | |
| | | | |
| | | | |
| | | | |
| | | | |
| | | | |
| | | | |
| | | | |
| | | | |
| | | | |